What's so FUNNY about Microbiology?

Joachim Czichos

Science Tech Publishers Madison, Wisconsin

Joachim Czichos, Kirchenbergweg 42,
D-6900 Heidelberg 25, Federal Republic of Germany

Science Tech Publishers, Inc., 701 Ridge Street
Madison, Wisconsin 53705 U.S.A.

Translation from German by Jane Ann Phillips.

Design and production by Science Tech Publishers.

Some of these cartoons originally appeared in issues of *forum mikrobiologie*.

ISBN 0-910239-12-6

Printed in the United States of America
10 9 8 7 6 5 4 3 2

Library of Congress Cataloging-in-Publication Data

Czichos, Joachim, 1955-
What's so funny about microbiology?

1. Microbiology -- Anecdotes, facetiae, satire, etc. I. Title.
QR56.C95 1987 576'.0207 87-12792
ISBN 0-910239-12-6

to Martina S., a phytopathology student who was the 'midwife' of my first cartoons.

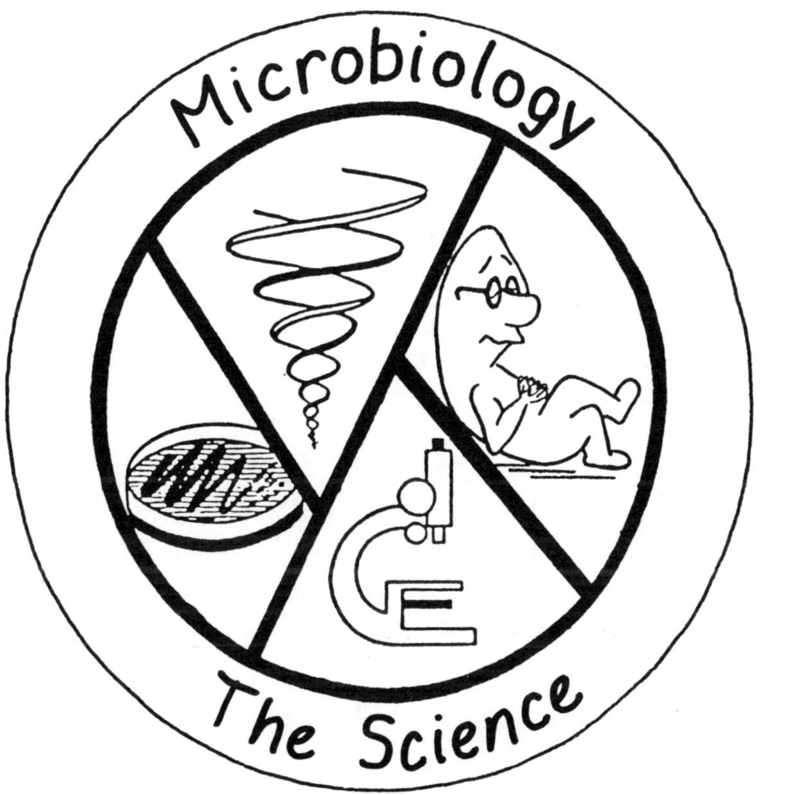
Microbiology
The Science

It is obvious that the branches of biology can be differentiated by their significance.
Micro-biology
Genetics
Nuclear Physics

For example, neither Systematics nor Paleontology has had any serious effect on many people, whereas research results in...
Micro-biology
Genetics
Nuclear Physics

... Molecular biology, Immunology, and of course Genetics are already changing the lives of all mankind.
Micro-biology
Genetics
Nuclear Physics

My distinguished colleague, you have overlooked Microbiology.
Micro-biology
Genetics
Nuclear Physics

My distinguished colleagues! I now welcome Professor Erich Coli — who will describe the essantial nature of microbiology.

You know, sometimes microbiology is absolutely incomprehensible to me.

The Microbiologist:

the life and times thereof...

Microbiologists come in two sizes – impressive and not-so-impressive.

They come in various shapes – oblong, spherical, and irregular.

Normally, microbiologists lead a solitary life, but sometimes they do appear in groups.

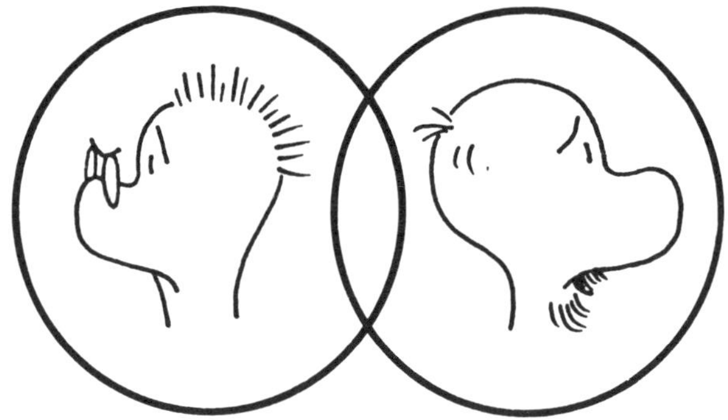

Microbiologists are, as a rule, peritrichously or monopolarly flagellated.

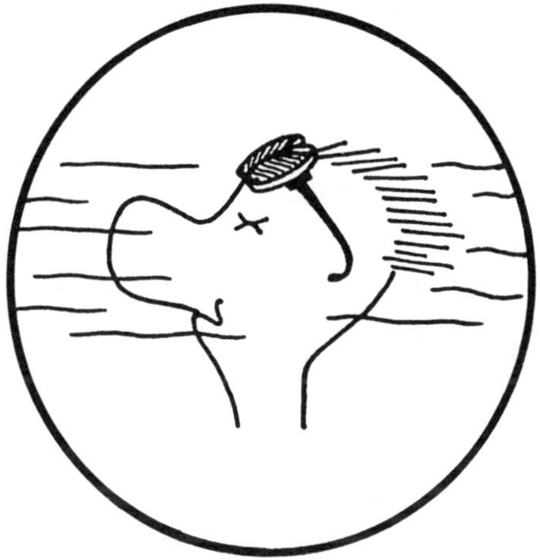

Generally, their lifestyle is obligately aerobic.

However, some species are microaerophilic.

The nutritional requirements of the microbiologist are very complex.

Although sterility is the rule in the lab, there is strong evidence for sexual propagation.

The work of an enzymologist begins with the preparation of a cell-free extract.

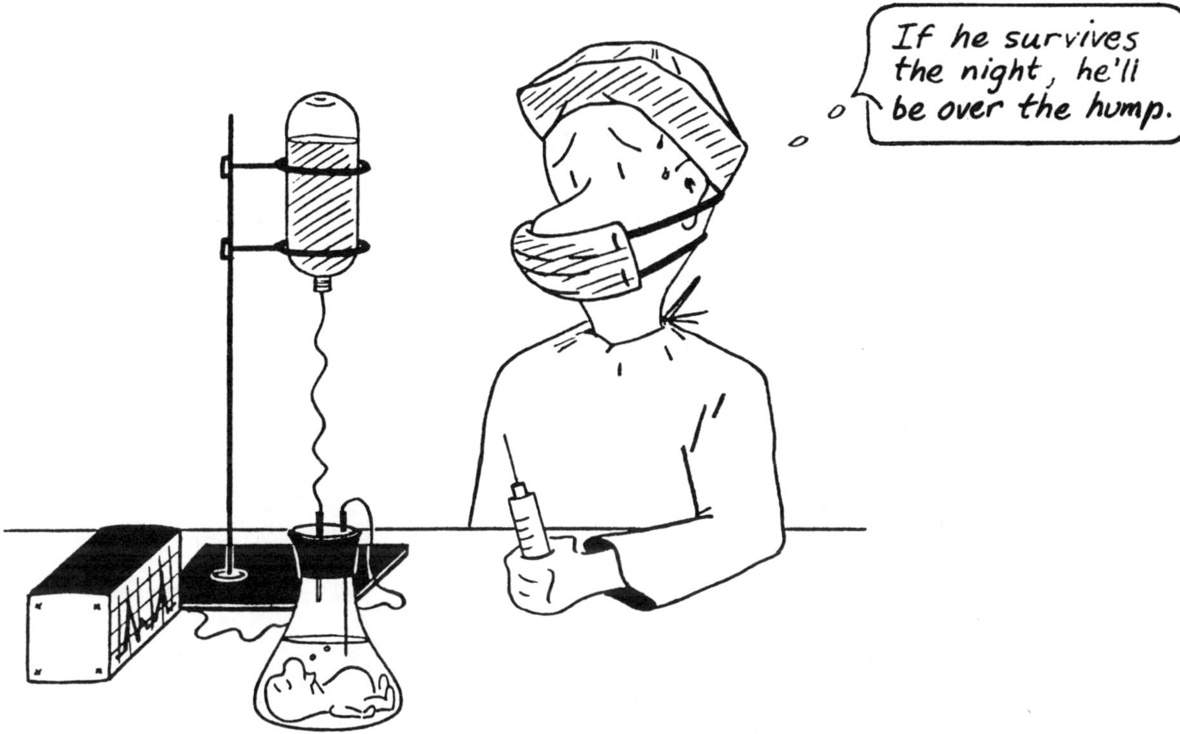

The medical microbiologist even cares about the life and death of microorganisms.

Food microbiologists don't like many kinds of microbes.

The skilled epidemiologist has to be a good detective.

The work of a microbial ecologist requires field work in distant lands.

Even the taxonomist has moments of glory.

An Institute
for Microbiology

Successful collaboration in an institute depends on a good working relationship between macro- and microorganisms.

A microbiology institute may be small in comparison to the large number of collaborators.

But even with its small size, scientists manage to keep the secretary busy.

Affiliates of the Institute
may be phototrophs.

The smallest affiliates
believe this is a
merry-go-round.

Some guests are unwelcome, but other visitors are always good to see... well, almost always.

An elevator can rapidly acclimate the new collaborator to how an institute works.

A library is a must!
For the books and sofa.

A security officer and his fearless watchdog keep the institute safe and sound...

So that research can
reach its ultimate goal.

The Microbiologist's greatest fear

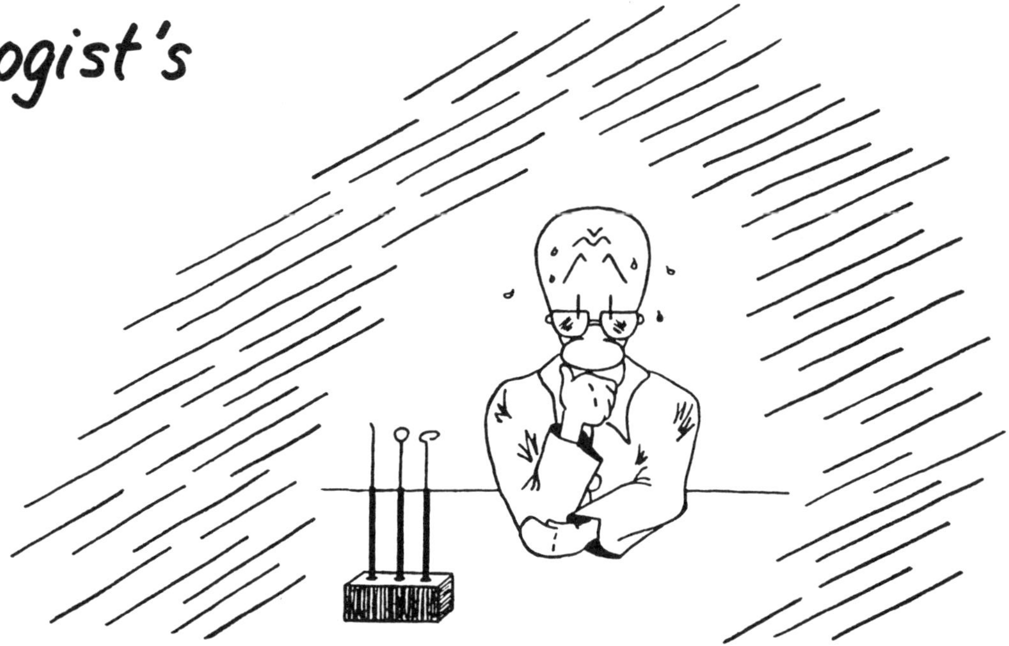

The microbiologist lives in constant fear of contamination, which threatens all pure cultures.
Air-borne contaminants
Pure cultures, boring but necessary

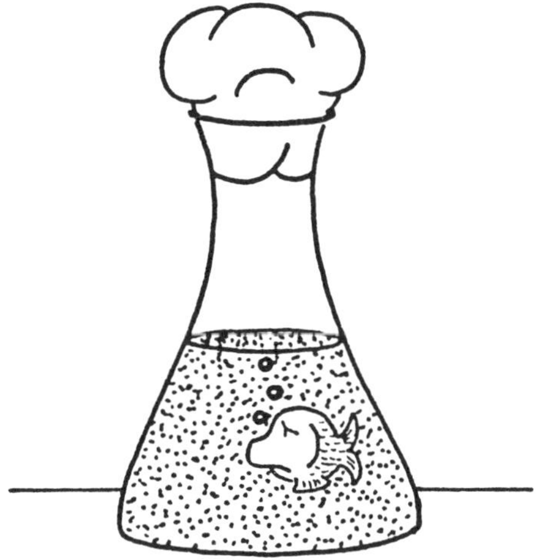

Contamination by eucaryotic organisms
can often be seen with the naked eye.

This deep-seated fear of contamination causes the microbiologist to ruthlessly perform mass sterilizations on these unwanted yet resigned microorganisms.

Some sterilization procedures are a little less ruthless.

Sterile material presents a challenge many microbes can't resist.

Even after all the precautions to prevent contamination, the conscientious microbiologist will still have nightmares at home.

On the road
to culture
In vitro
In vivo

Bacteria in their natural environment, living happily together.

And then along comes a microbiologist.

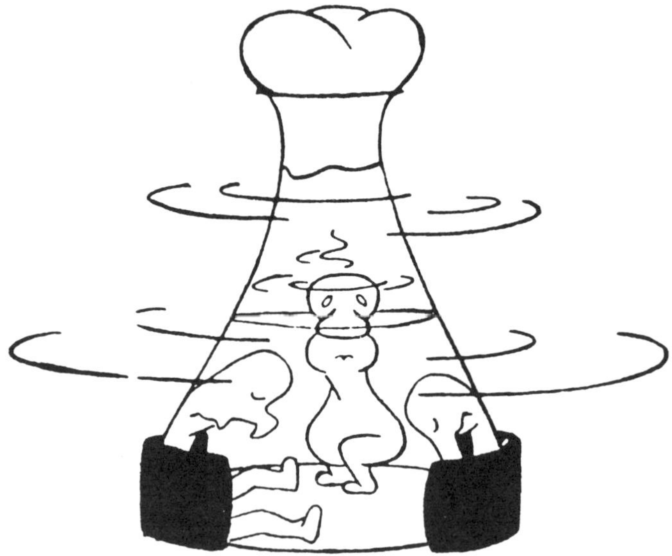

Only a few bacteria withstand the rigors of enrichment culture in laboratory media.

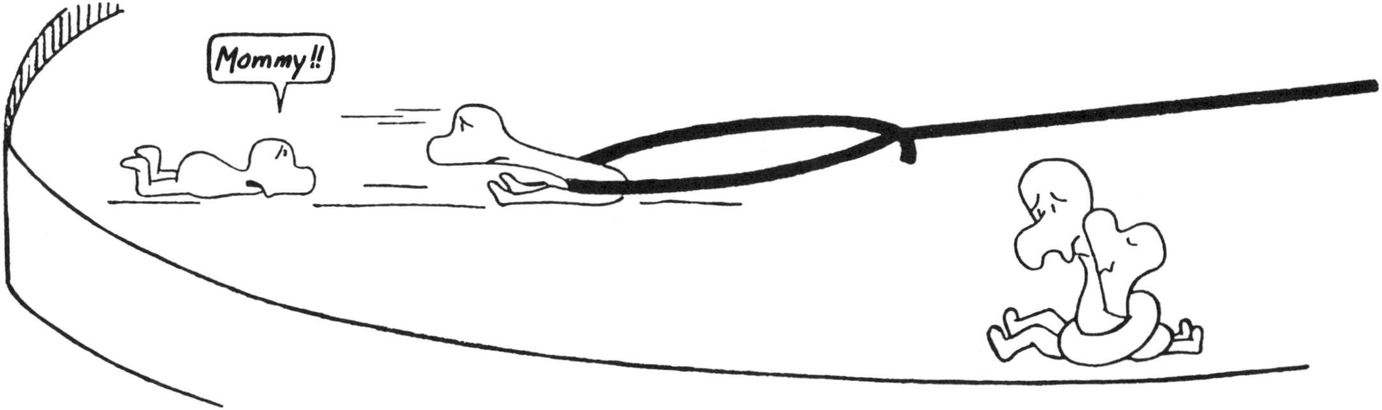

Torn from their families,
a few poor, pathetic souls
are isolated onto agar plates.

However, sometimes
there are fast familial
ties that cannot be
broken.

The isolated bacteria give rise to single colonies of genetically identical, listless individuals.

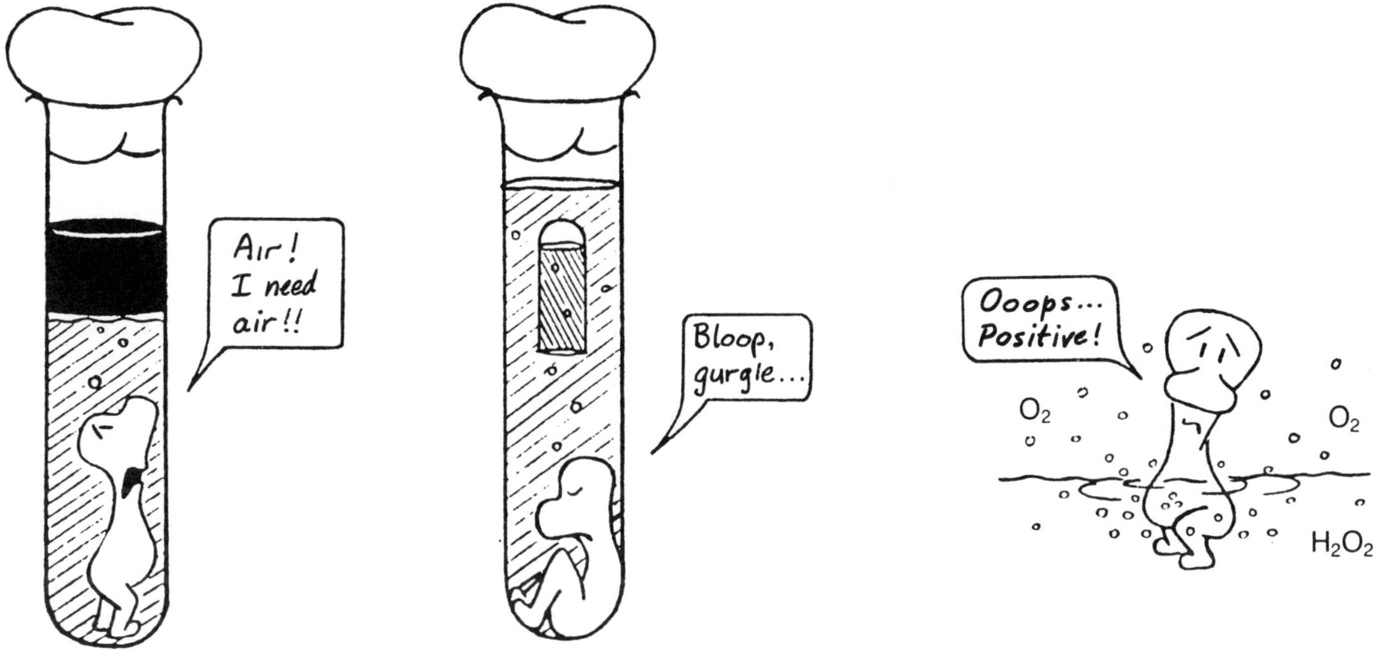

Pure cultures of bacteria are subjected to a variety of morphological, physiological, and biochemical tests.

Bacteria that can endure all the tests for classification are finally given names.

Life outside the laboratory

Despite the increasing number of microbiologists, there will always be bacteria which will remain unknown.

These bacteria live in constant fear of being isolated.

All the best places to live
are already occupied.

The colonization of a new microniche remains, for most of them, a dream.

Homeless, trusting in the wind and water, these bacteria have still been successful.

Some are even found where no creature in its right mind should live.

A despairing few surrender their fates to the microbiologist....

but others trust in a better future →

We'll always survive, no matter what!

Wild type and mutant bacteria

The wild-type microorganism lives a
hardy life in its natural environment.

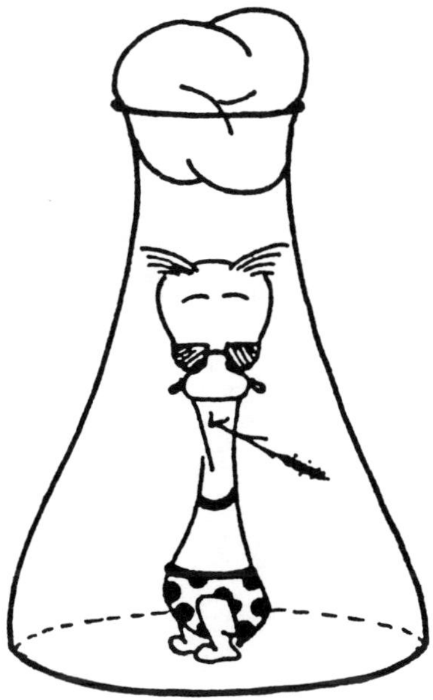

After isolation and many generations in the laboratory, there is a striking change in the microorganism's characteristics.

During the microbiologist's search for mutants, the appearance degenerates even more.

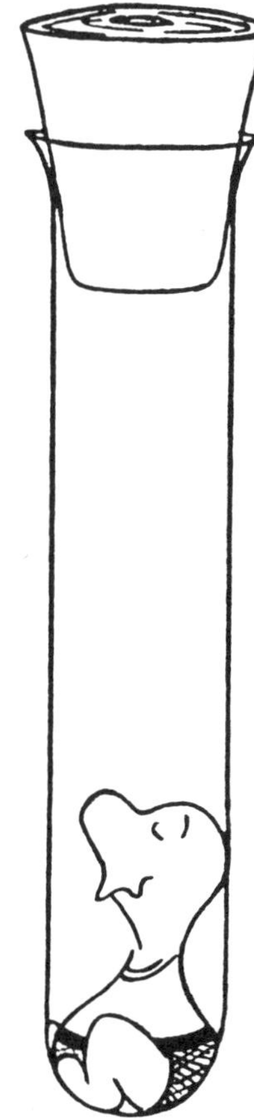

Ultraviolet light is used to induce many mutations, although some bacteria do not seem to mind this treatment.

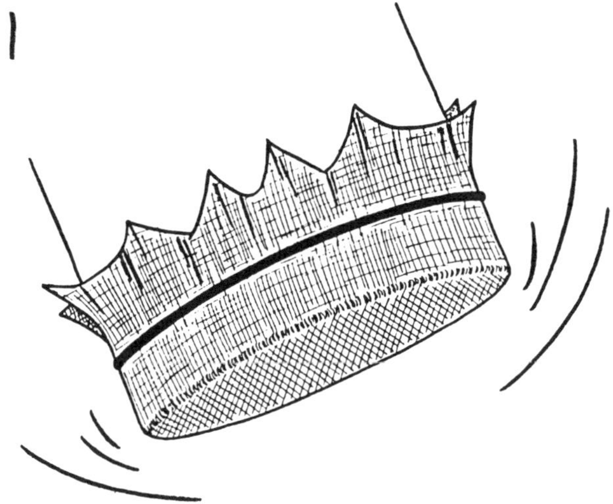

Through Lederberg's replica plating technique, many different kinds of selective pressure can be applied.

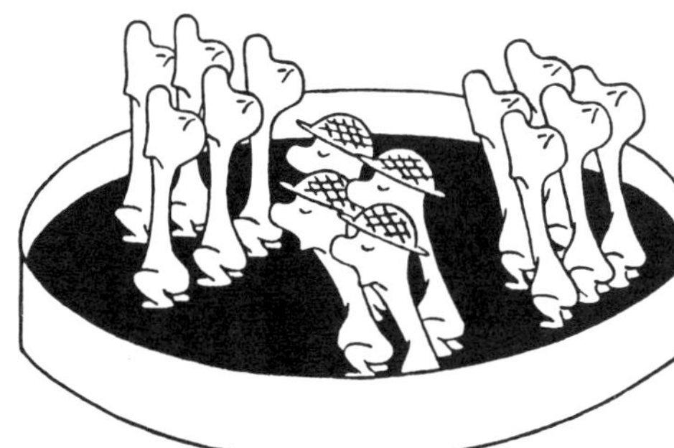

Among the various mutants, one can find the desired phenotype,

various defective mutants,

and, of course, revertants.

It's no wonder after all this that some microbes choose to escape from the laboratory.

Bacterial lifestyles

Bacteria can be found that have many different life styles.

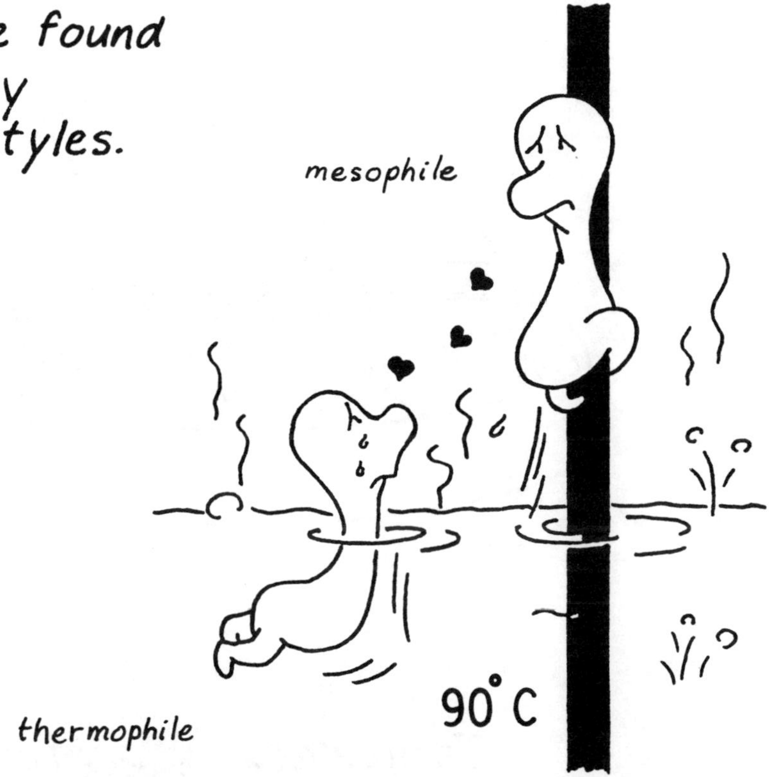

Temperature preferences.

pH preferences

Preferences for other environmental conditions.

aerophile

halophile
20% NaCl

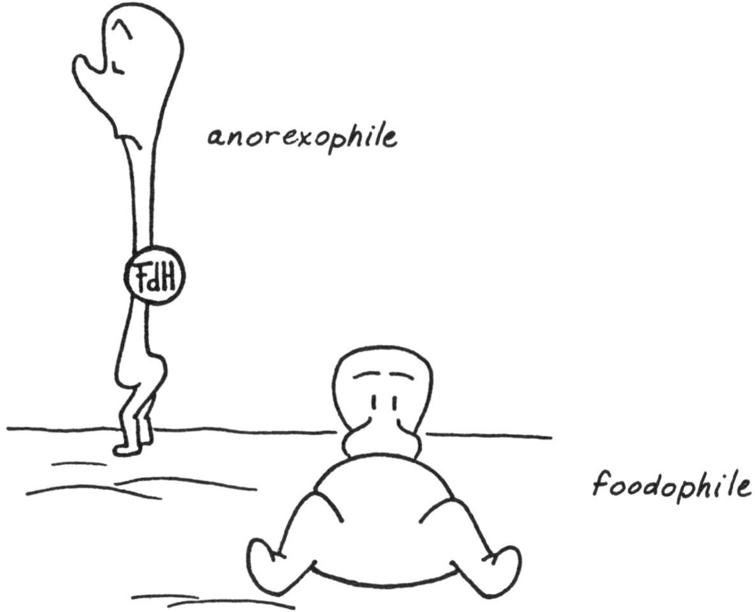

Nutritional preferences.

And, of course, the bacterial type found in all microbiology laboratories.

Function of the bacterial cell envelope

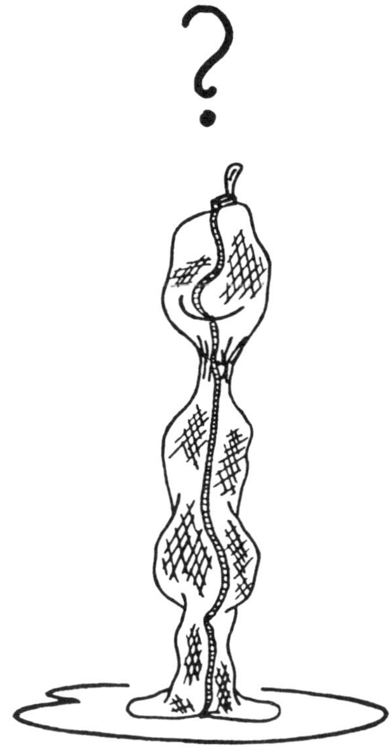

Specific cell-to-cell contact

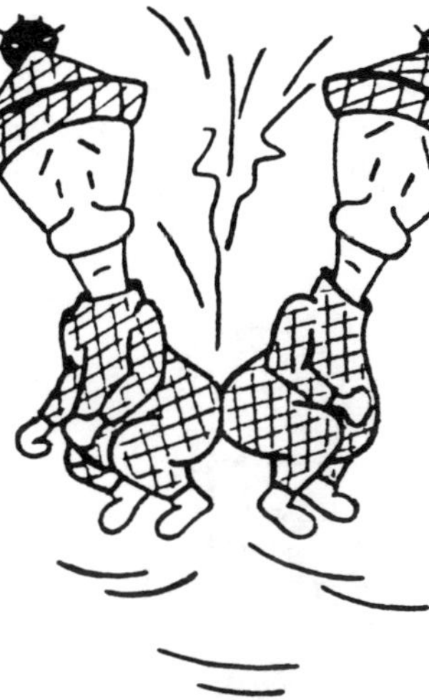

Homologous cells

Heterologous cells

Nonspecific adhesion

Molecular sieve

Phage receptors

Protection

Cation binding

Shape

I guess I won't need that in here.
In vitro

And I wonder what the microbes think —

How nice to be an autotroph... sometimes.

Phototrophy must be the life!
heterotroph
H_2COH
$h \cdot \nu$
heterotroph
H_2COH
OH
OH
OH
OH
phototroph
heterotroph
H_2S

Gas vacuole
Poor guy- he has gas again!

Sic 'em!
β-Lactamase
Penicillin

Sorry kid, my mom said not to let anyone in to play.
E. coli

H_2S
H_2S
H_2S
H_2S
H_2S
My deodorant didn't work.
H_2S
H_2S
H_2S

Ever heard about conjugation, honey?

Fungi,

and other eucaryotes

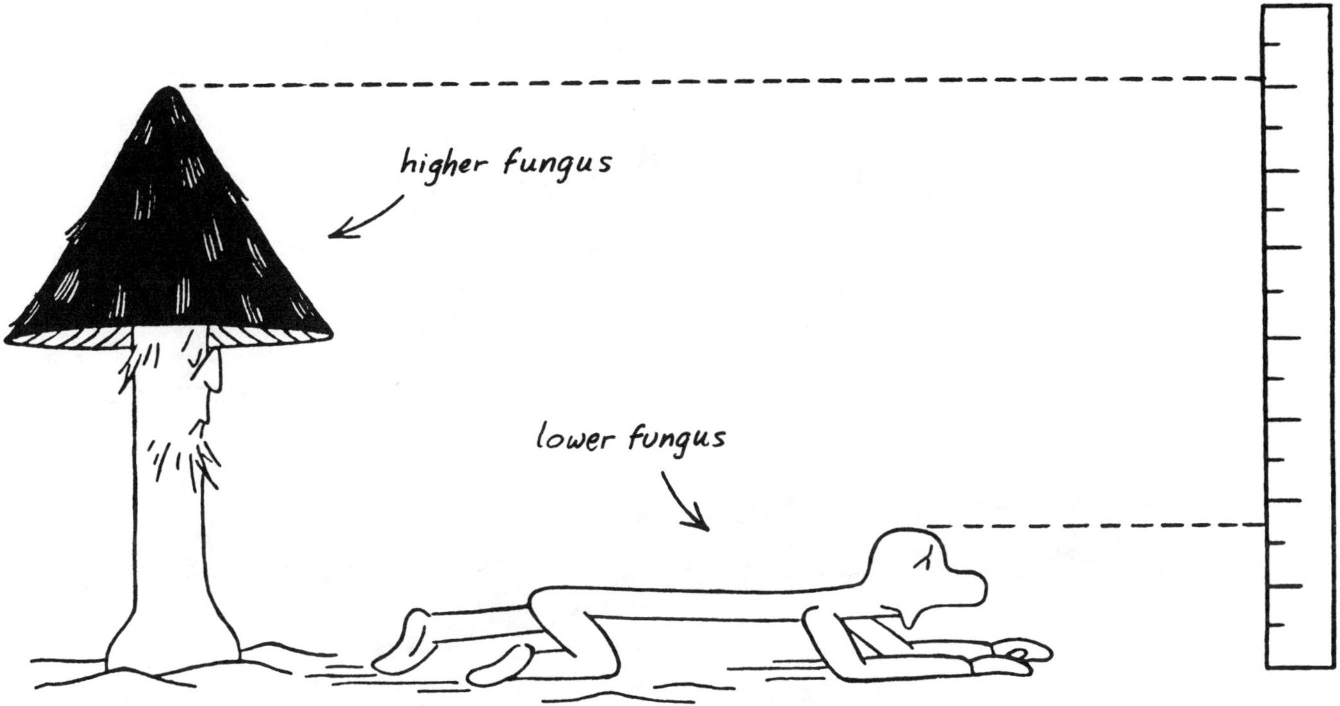

There are higher and lower fungi.

The Fungi imperfe~~k~~cti resist orderly classification.

The slime molds seem to have an incredible urge to be sociable.

Dictyostelium — the unicellular amoeboid cells ...

... Aggregate
to form a
multicellular
fruiting body.

Yeast cells have both sexual and asexual modes of reproduction.

Algae are sun worshippers. Fungi wish they were.

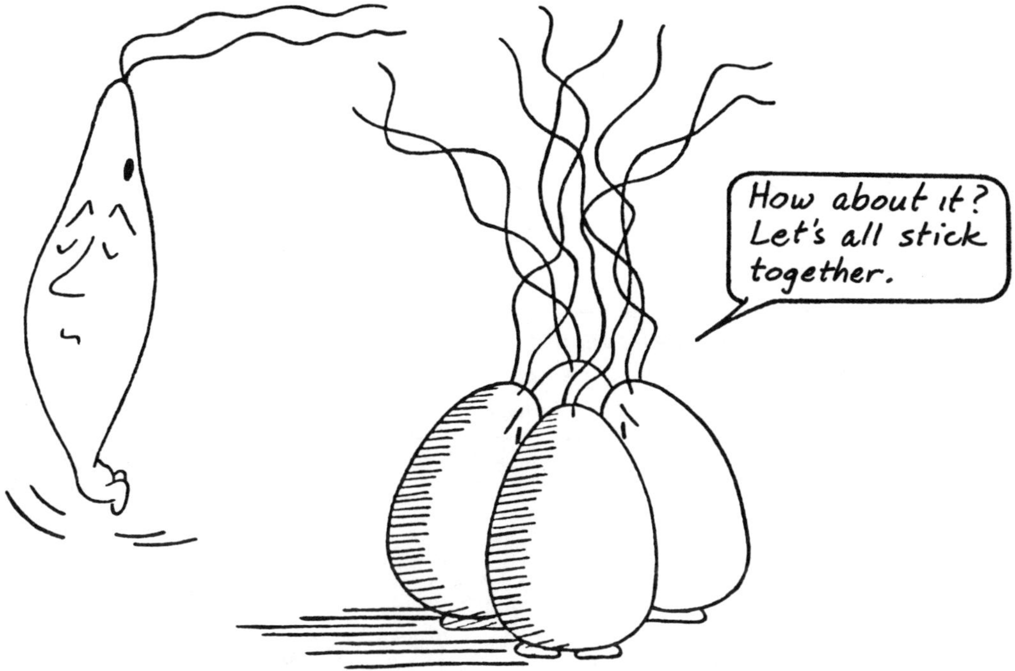

There are all kinds of green algae. Some are loners, others are more social.

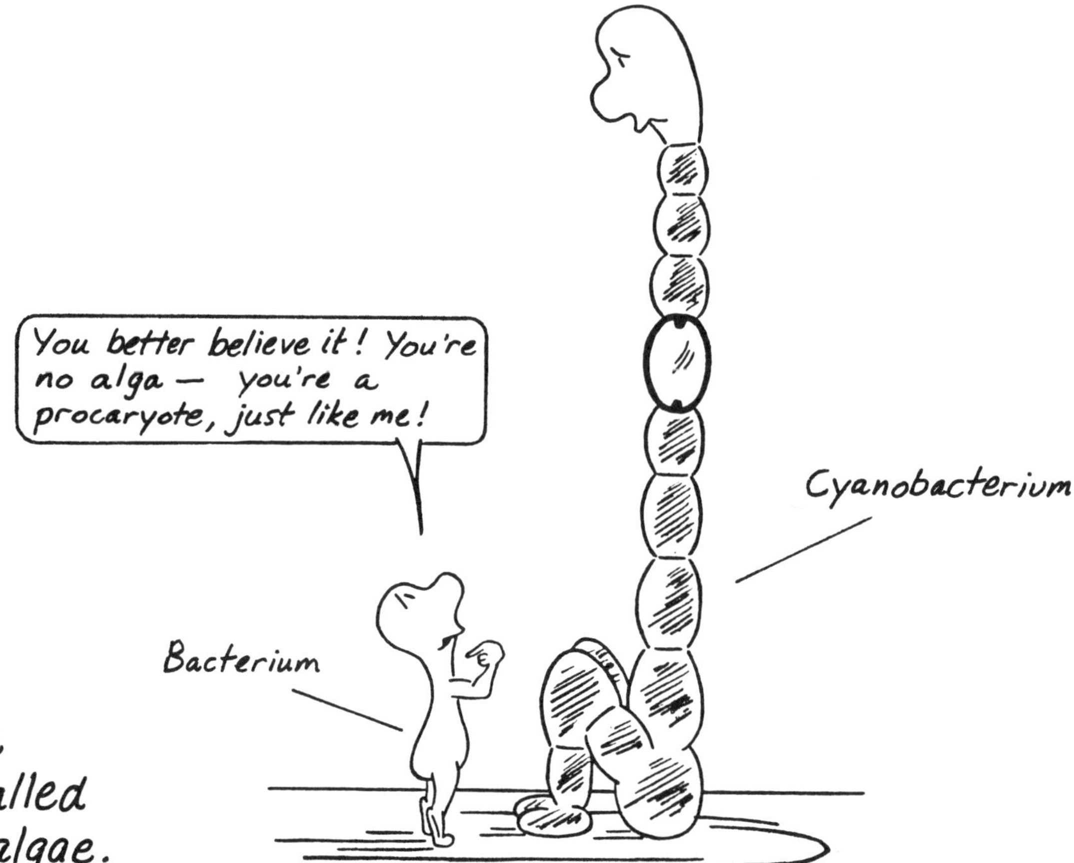

Cyanobacteria used to be called Blue-green algae.

Some amoebae have houses; others wander the world unclothed and homeless.

The variety of forms among the flagellates and the ciliates is simply astounding.

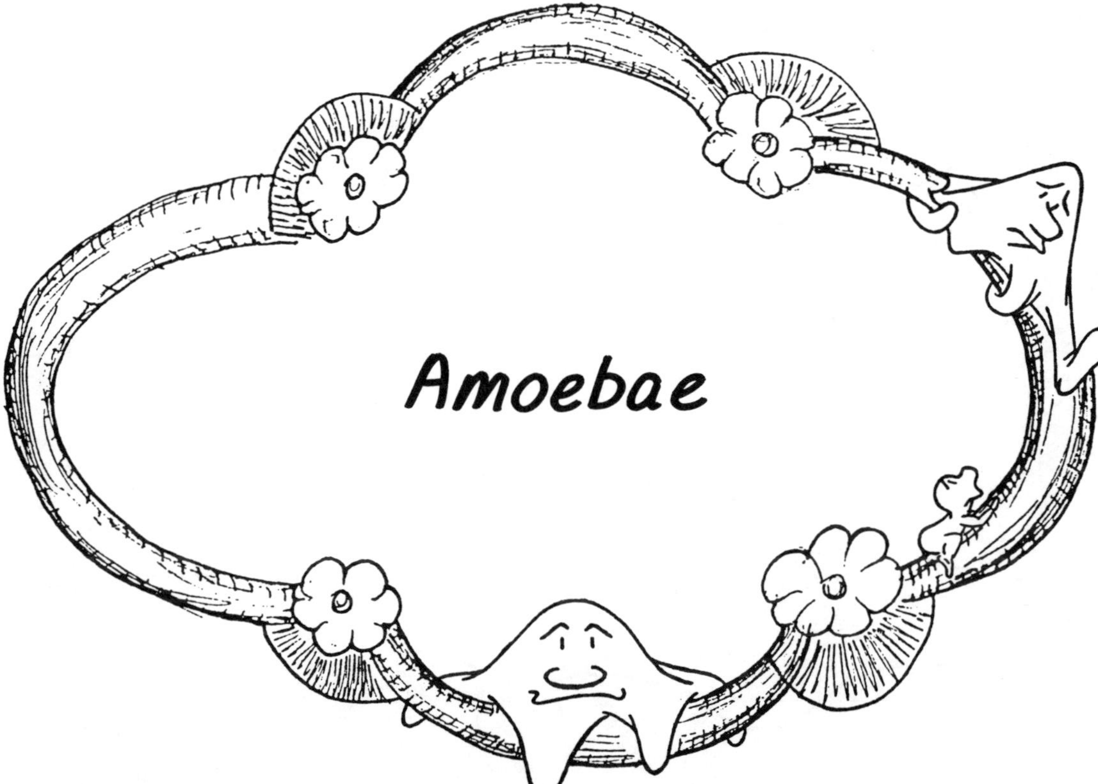

Amoebae

Amoebae are eucaryotes,

shapeless,

and frighteningly **BIG.**

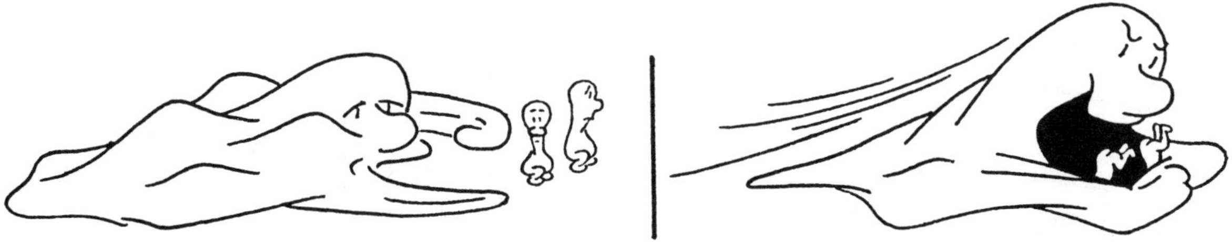

Amoebae can be sneaky,

cruel,

and seem to be
most insistent.

Amoebae sometimes seem to be quite cowardly....

....and an experienced worker can keep them under control.

Sometimes they're not too bright.

The study of Pathogenesis

An apparently insignificant contact between a pathogen and its host cell already seals the fate of the unfortunate host cell.

Specific recognition and attachment to host cell receptors is indispensable to the bacteriophage.

In animals, a showdown with the immune system may finally decide if penetration is to be.... or not to be.

In plants, with no immune system, each cell must be ready to fight.

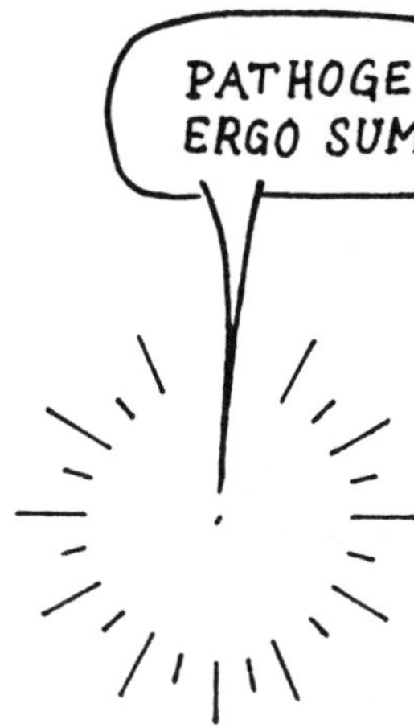

Viruses are so small that they can hardly be studied!

Even microbes are not safe from each other.

And another fateful encounter.

But hosts are not alone!
The ambitious microbiologist
is hot on the trail
of pathogens.

And he's found that for many pathogenic bacteria, the reasons for their pathogenicity are obvious.

But others have not yet surrendered the secrets of their pathogenicity.

At least one reason for the secrecy is that many virulent pathogens become completely innocuous after being in laboratory culture.

Unfortunately, the same can sometimes be said of the ambitious microbiologist.

Caution!

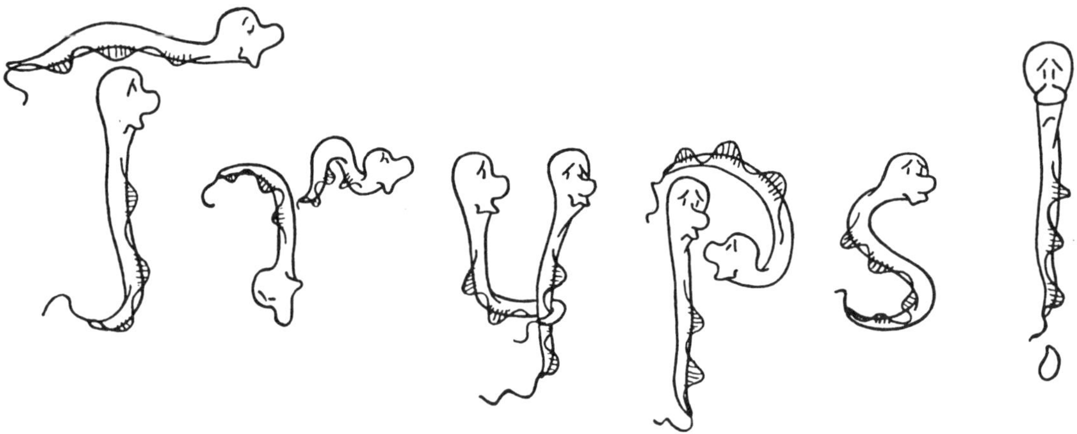

Trypanosomes which look completely harmless ...

are actually pathogens of man and animals.

The trypanosomes are so successful because of frequent changes of their antigenic coats.

Therefore, an extensive wardrobe of heterogeneous coat types is indispensable equipment for them.

When a tsetse fly takes a blood meal from an infected animal, the ingested trypanosomes rapidly shed their coats.

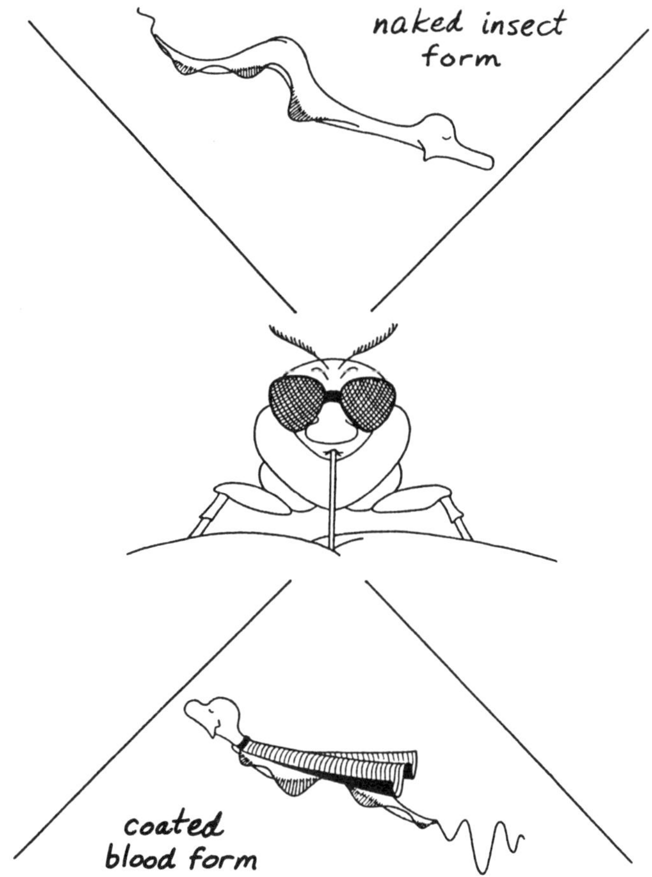

Injected back into their host's blood, these causative agents of sleeping sickness are ruthless.

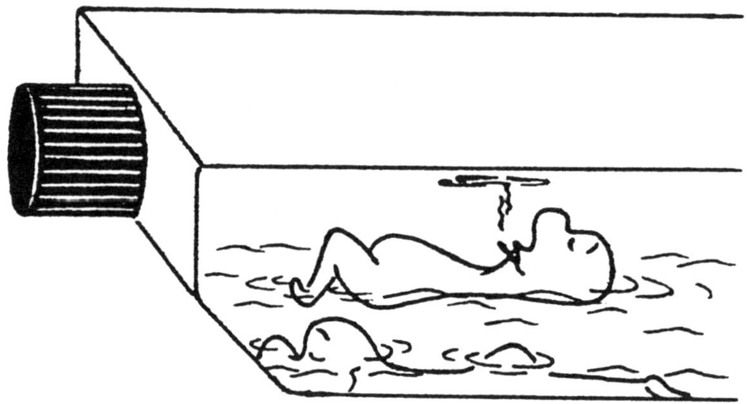

But many lead a more cultured
life in research laboratories.

Recent work has uncovered some evidence of a sexual cycle in trypanosomes.

Investigations of the shedding of the surface coat may lead to successful chemotherapy because naked trypanosomes are destroyed by the host's immune defense.

Transcription description

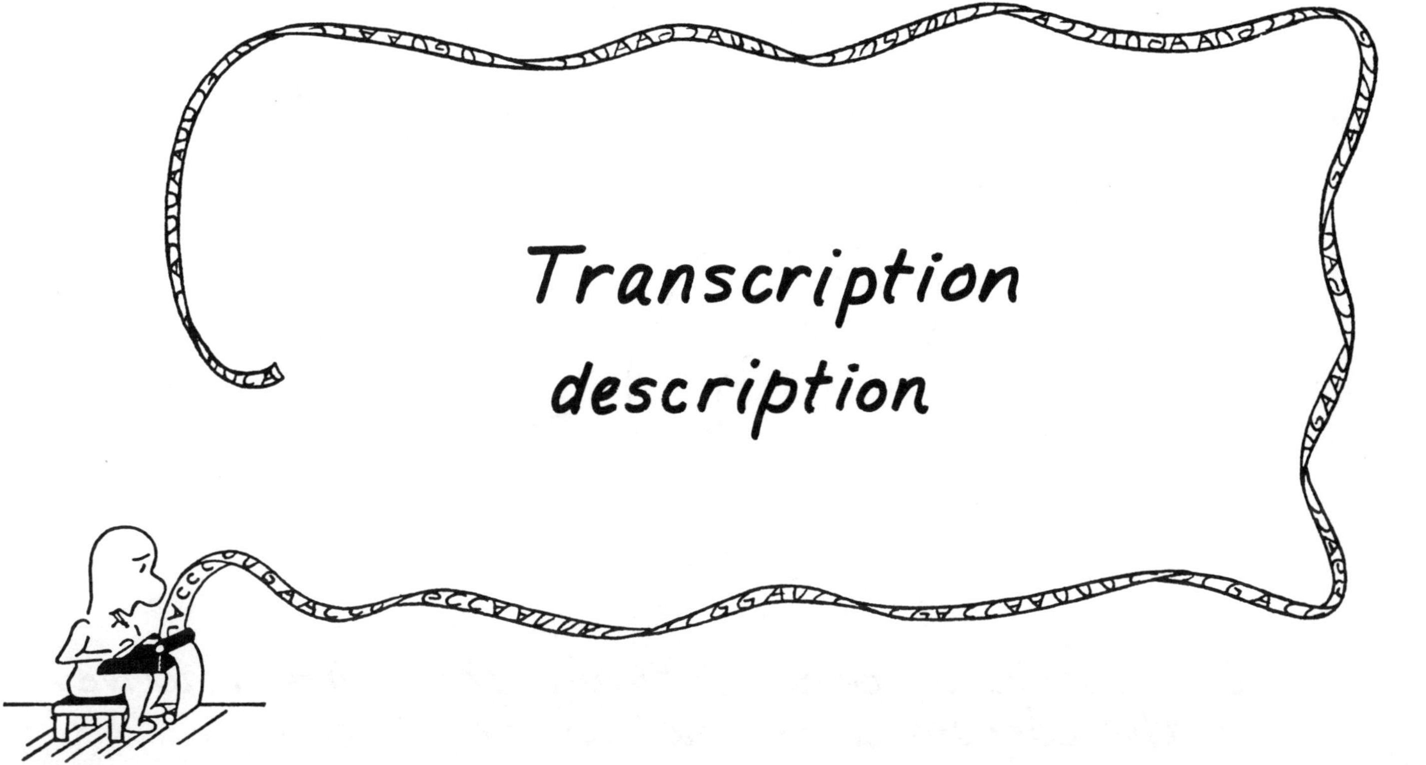

In eucaryotic cells, dormant genes are inactive in the condensed chromatin structure.

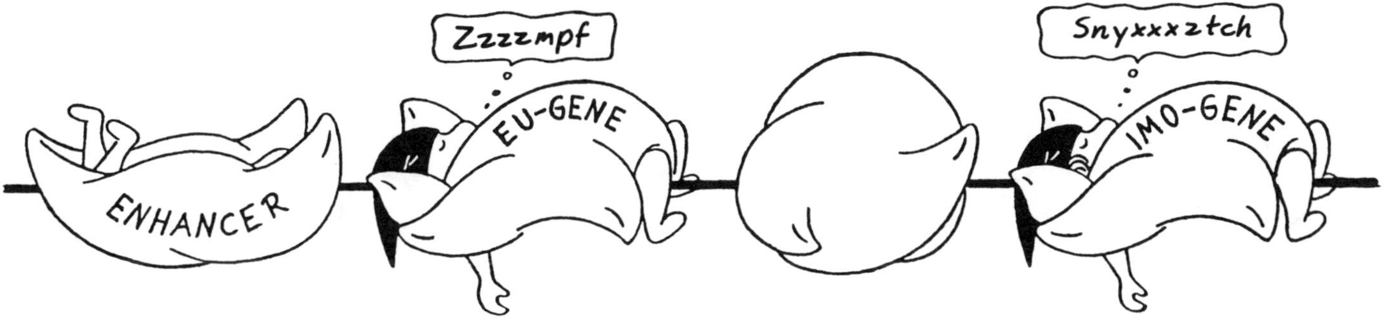

Gene activation requires unfolding of a local region of the chromosome.

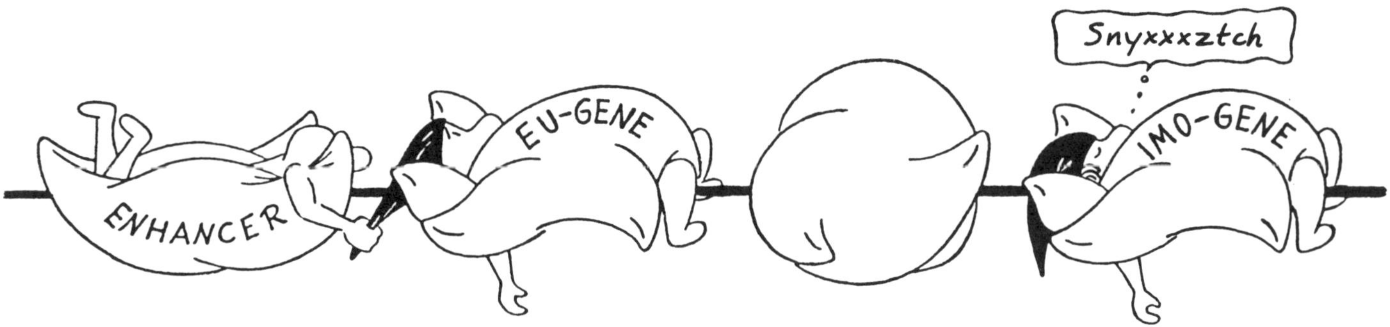

The gene is converted into an active form by cis-acting DNA elements and trans-activating protein factors.

The enhancer puts the gene in position....

so that the trans-acting factors are more likely to bind to the gene.

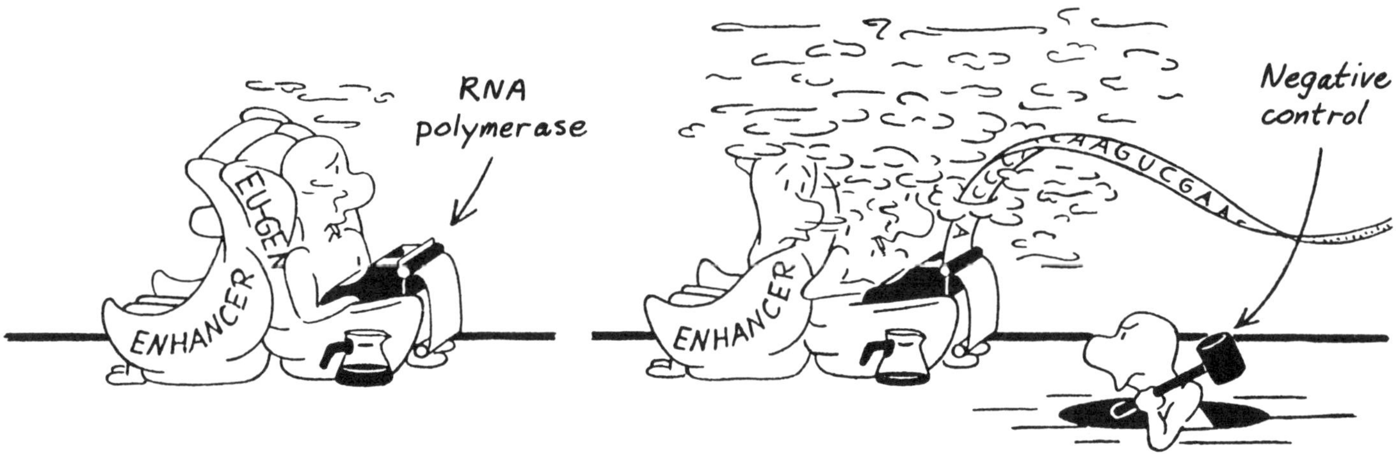

After binding of RNA polymerase, RNA is transcribed from the fully activated gene until it is shut off by the action of negative control elements.

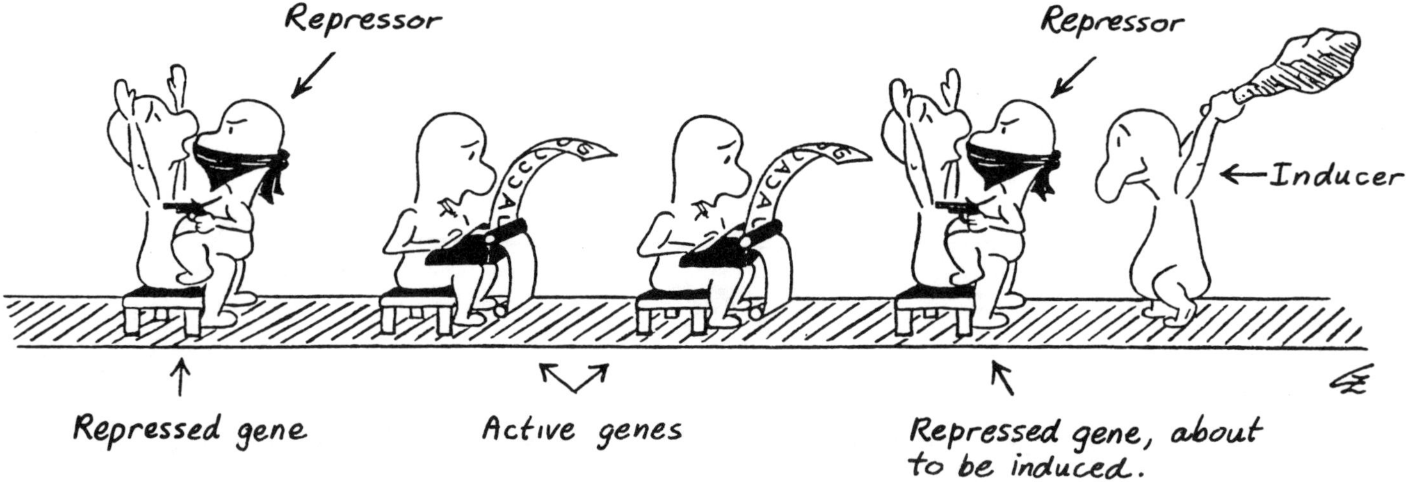

The genes of procaryotic cells are not allowed to sleep. Compared to eucaryotic cells, their activity is regulated by methods which are rather crude, but effective.

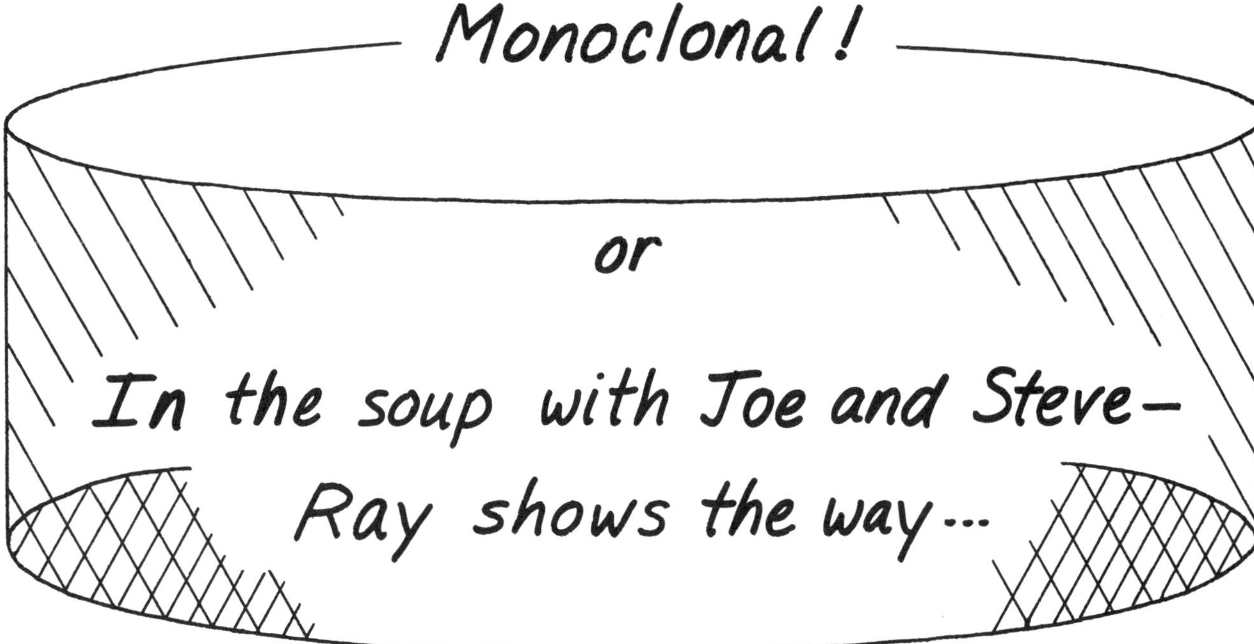
Monoclonal!
or
In the soup with Joe and Steve–
Ray shows the way...

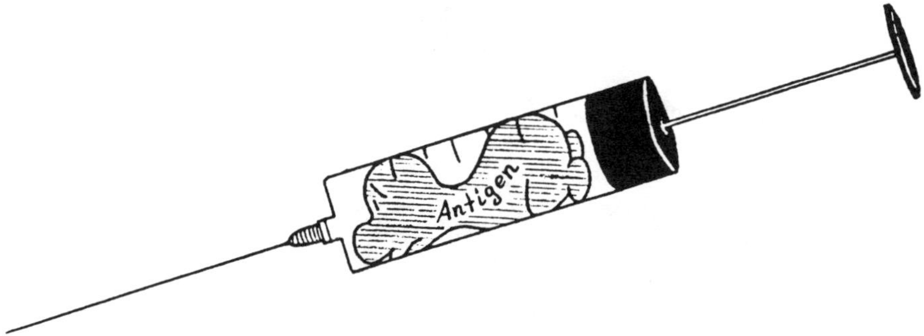

Here comes the antigen, dissolved in buffer, mixed with adjuvants, ready to be injected into an experimental animal. The antigen has a last moment of peace before ....

(Not to scale!)

Bumping into some resting, waiting B-lymphocytes. And then...

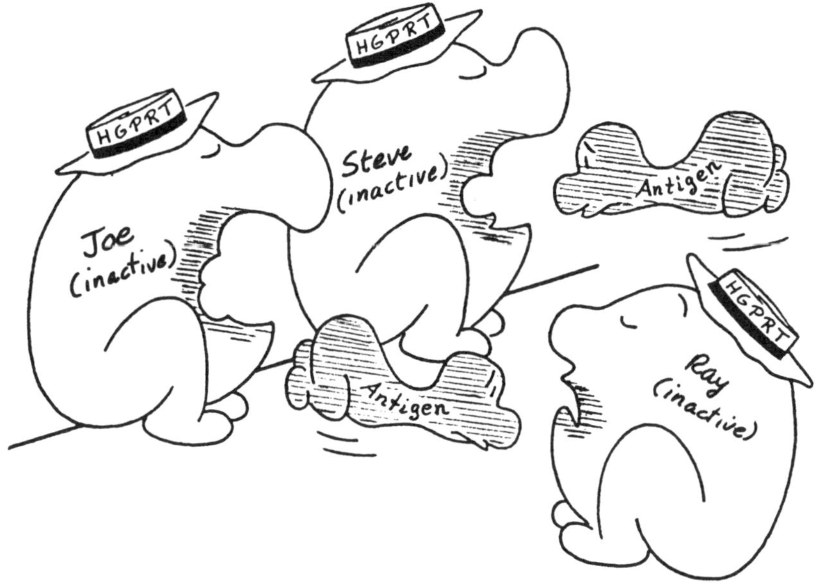

HGPRT, the enzyme necessary for growth in HAT medium. What is HAT? Not a chapeau, but... Read on...

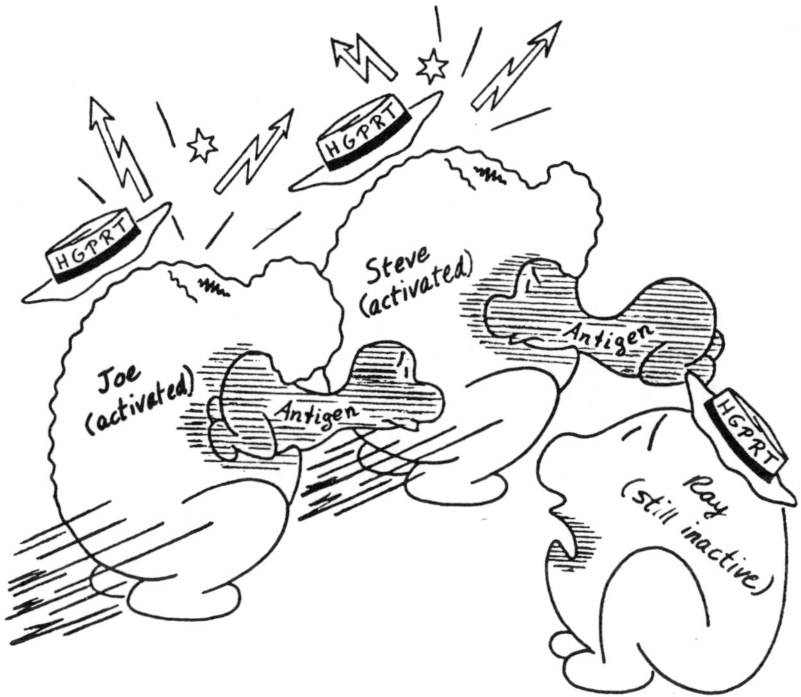

Activation!! Joe and Steve will never be the same again. Ray, the lazy bum, still rests.

A meeting is arranged with the brothers Grimm, myeloma cells, which are immortal.

Immortality
(Grows forever in culture)

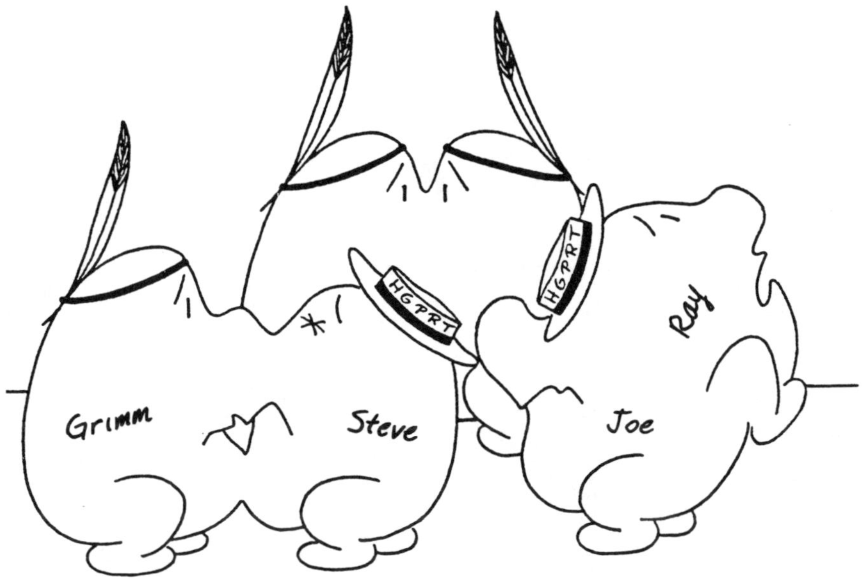

Cell fusion! Some of the fused cells are antibody-forming hybridoma cells.

Using HAT selection in a special medium, the hybridoma cells selectively reproduce and are cloned. The other cells retire.

The sought-for hybrid clone forms only one kind of antibody — a "monoclonal" antibody.

Biotechnology

Decisions! Decisions!
c-DNA Bank
Special
Plasmid salad with $CaCl_2$ Dressing

He carries a human gene, but nobody knows exactly which one!
Sure! And I get stuck with a chicken's gene!

... and ever since my transformation, ampicillin's no problem!
AmP

Overproduction: the incontinent microbe

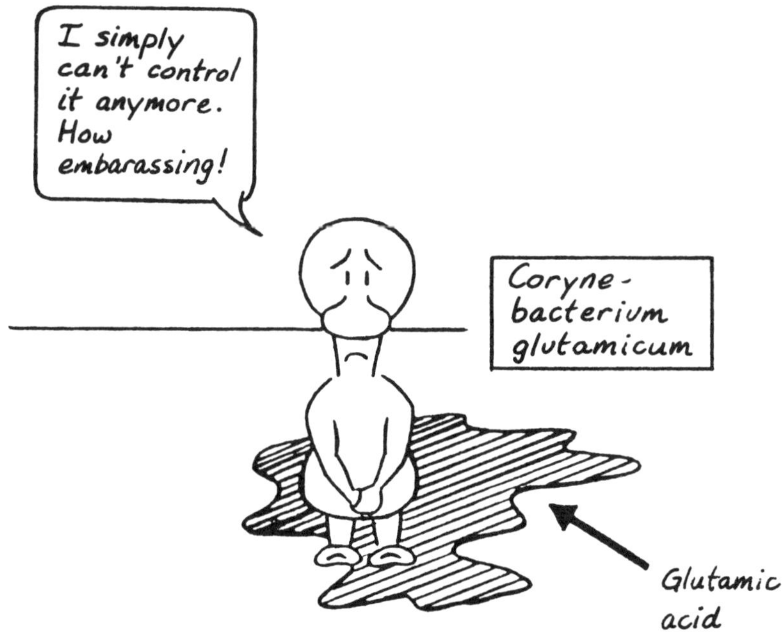

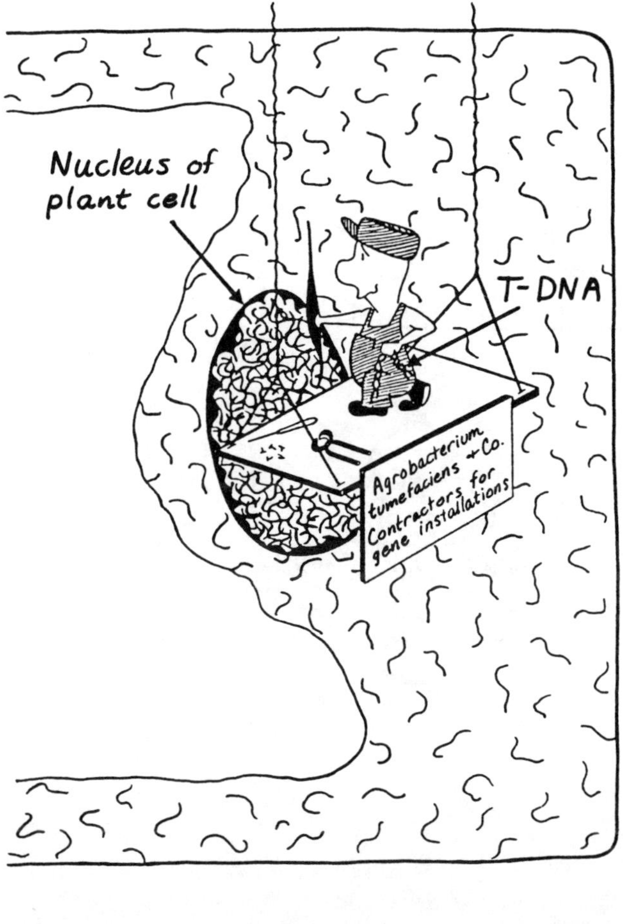
Nucleus of plant cell
T-DNA
Agrobacterium tumefaciens + Co.
Contractors for gene installations

Yesterday, I didn't even know what insulin was!
Ins
Insulin
sulin
Insulin
Personally produced for you

Oh sure! When the beer is finally ready to drink, they throw us away!
Saccharomyces cerevisiae

And a few more glimpses of bacterial life

In an anaerobic digestor

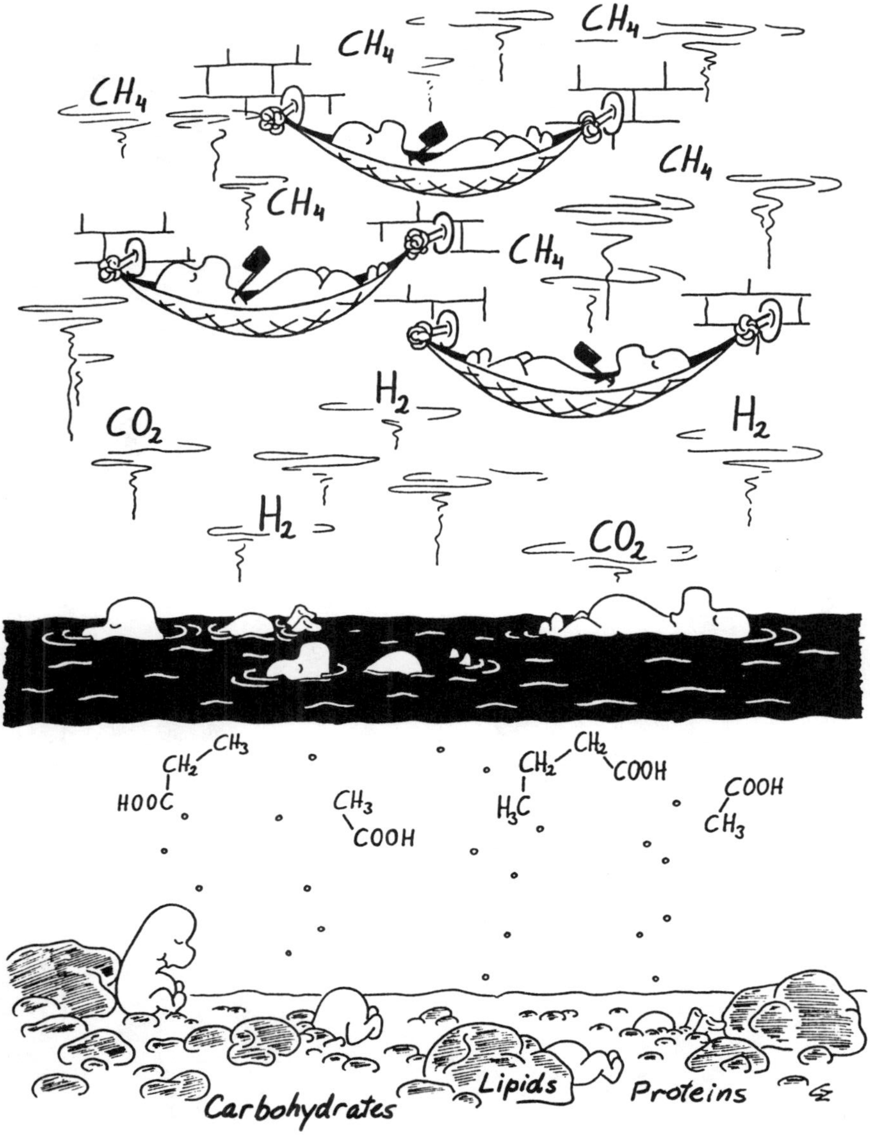

In the mines — mechanisms possibly used by microorganisms in the decomposition of coal.

Incorporation of whole coal particles

Uptake of components which are soluble in the medium

Decomposition of coal by free extracellular enzymes

Decomposition of coal constituents by extracellular cell-wall-fixed enzymes

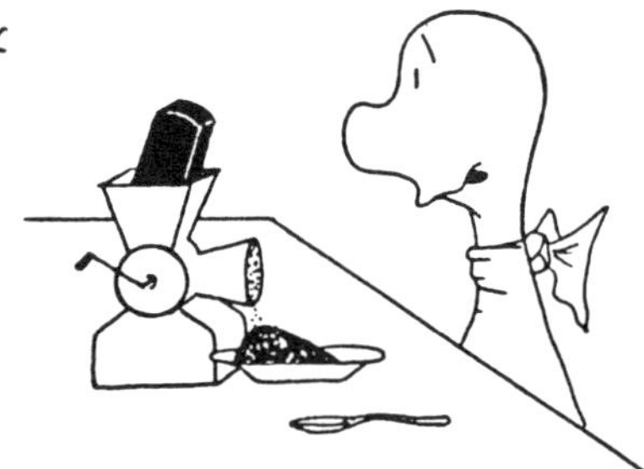

And on mine tailings

And on the farm

In the laboratory

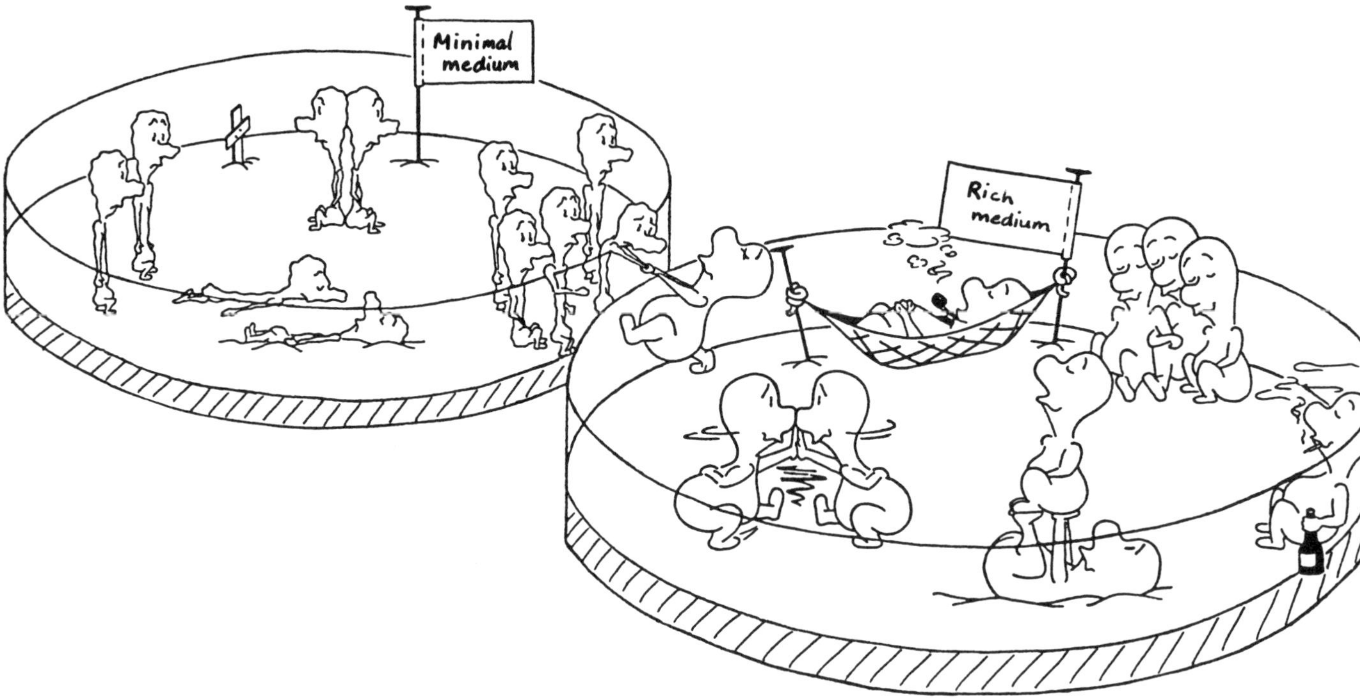
Minimal medium
Rich medium

And on vacation (Salt Lake City, of course).
A phototrophic halophile
20% NaCl

And who
will have
the last word?

"It is microbes who will have the last word."